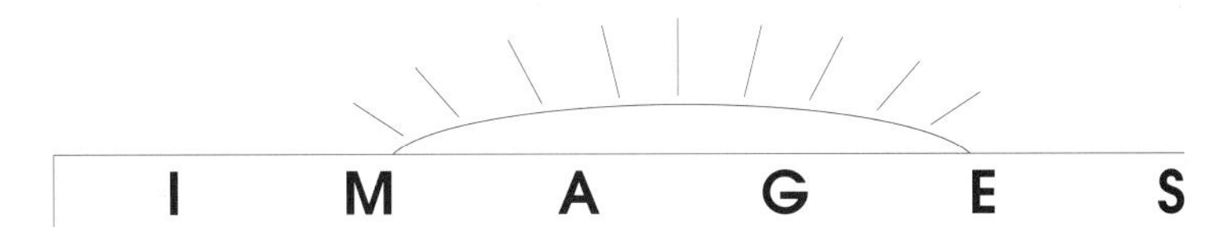

# In the Town

Karen Bryant-Mole

First published in Great Britain by Heinemann Library, Halley Court, Jordan Hill, Oxford OX2 8EJ,
a division of Reed Educational & Professional Publishing Ltd.

OXFORD FLORENCE PRAGUE MADRID ATHENS MELBOURNE AUCKLAND KUALA LUMPUR
SINGAPORE TOKYO IBADAN NAIROBI KAMPALA JOHANNESBURG GABORONE
PORTSMOUTH NH (USA) CHICAGO MEXICO CITY SAO PAULO



**Designed by Jean Wheeler**
**Commissioned photography by Zul Mukhida**

*Printed in Hong Kong / China*

01 00 99
10 9 8 7 6 5 4 3 2 1

*ISBN 0 431 06313 3*

British Library Cataloguing in Publication Data
Bryant-Mole, Karen
In the Town. - (Images)
1.Cities and towns - Juvenile literature
2.City and town life - Juvenile literature 3.Readers (Primary)
I.Title
307.7'6

**Some of the more difficult words in this book are explained in the glossary.**

**Acknowledgements**
The Publishers would like to thank the following for permission to reproduce photographs. Bruce Coleman Ltd; 12 (right) Jane Burton, 13 (right) Frieder Sauer, 16 (right), Chapel Studios; 4 (right), 20 (right) Zul Mukhida, James Davis Travel Photography; 11 (right), Oxford Scientific Films; 12 (left) Mike Birkhead, Tony Stone Images; 5 (top) Stephen Johnson, 5 (bottom) John Lamb, 10 (right) David Madison, 13 (left) Geoff Johnson, 16 (left) Don Smetzer, 17 (right) Nicole Katano, 20 (left) Wayne Eastep, 21 (right) Richard Passmore, Zefa; 4 (left), 10 (left), 11 (left), 17 (left), 21 (left).

Cover map; The infopoint Map Company

Every effort has been made to contact copyright holders of any material reproduced in this book. Any omissions will be rectified in subsequent printings if notice is given to the Publisher.

# Contents

Buildings 4
Shops 6
Food shops 8
Transport 10
Wildlife 12
Fast food 14
Play park 16
Signs 18
Places to visit 20
Collections 22
Glossary 24
Index 24

# Buildings

There are lots of buildings in towns.

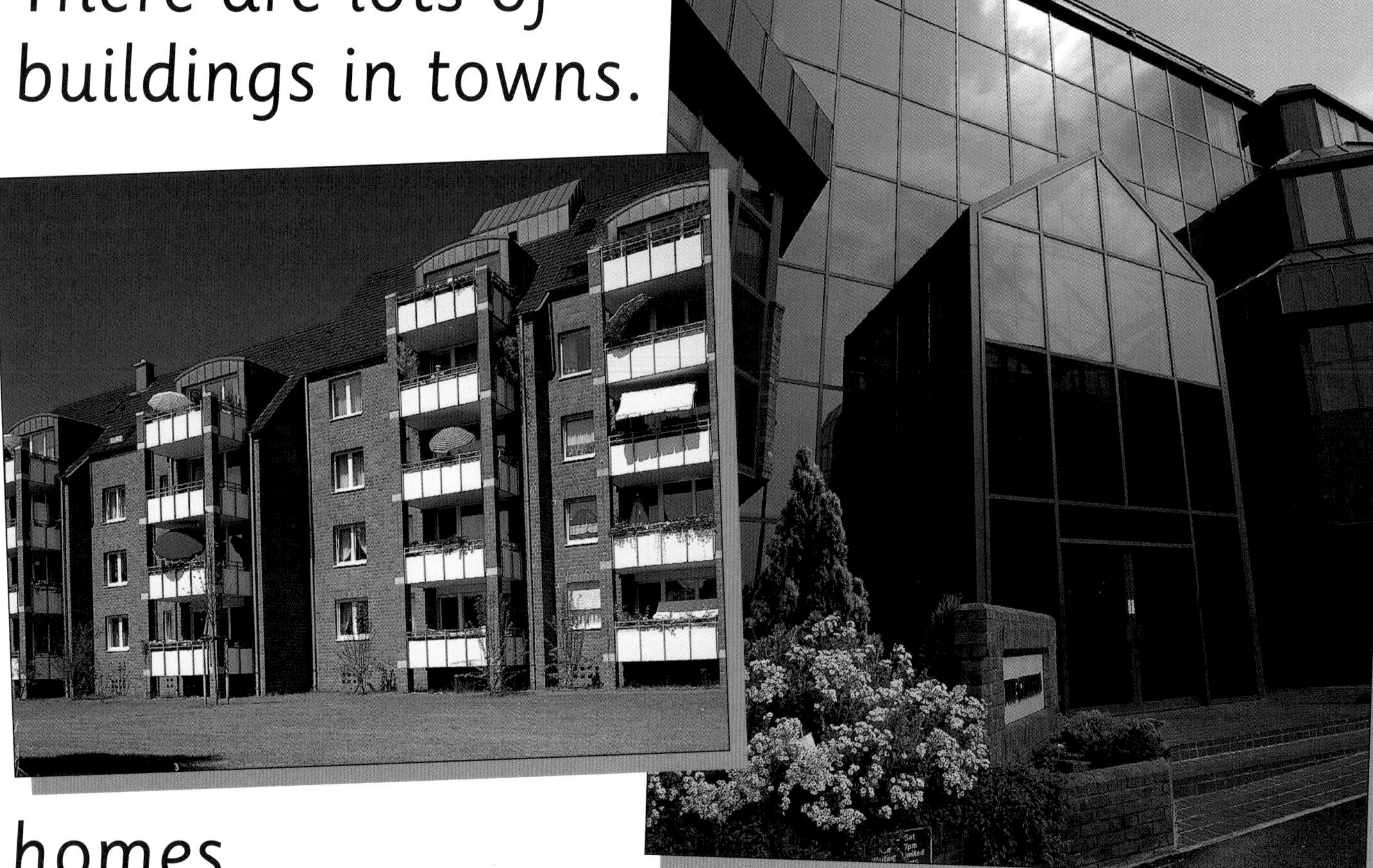

homes

office

Different buildings are used in different ways.

shops

restaurant

# Shops

You can buy all sorts of things in the shops in towns.

shoes

flowers

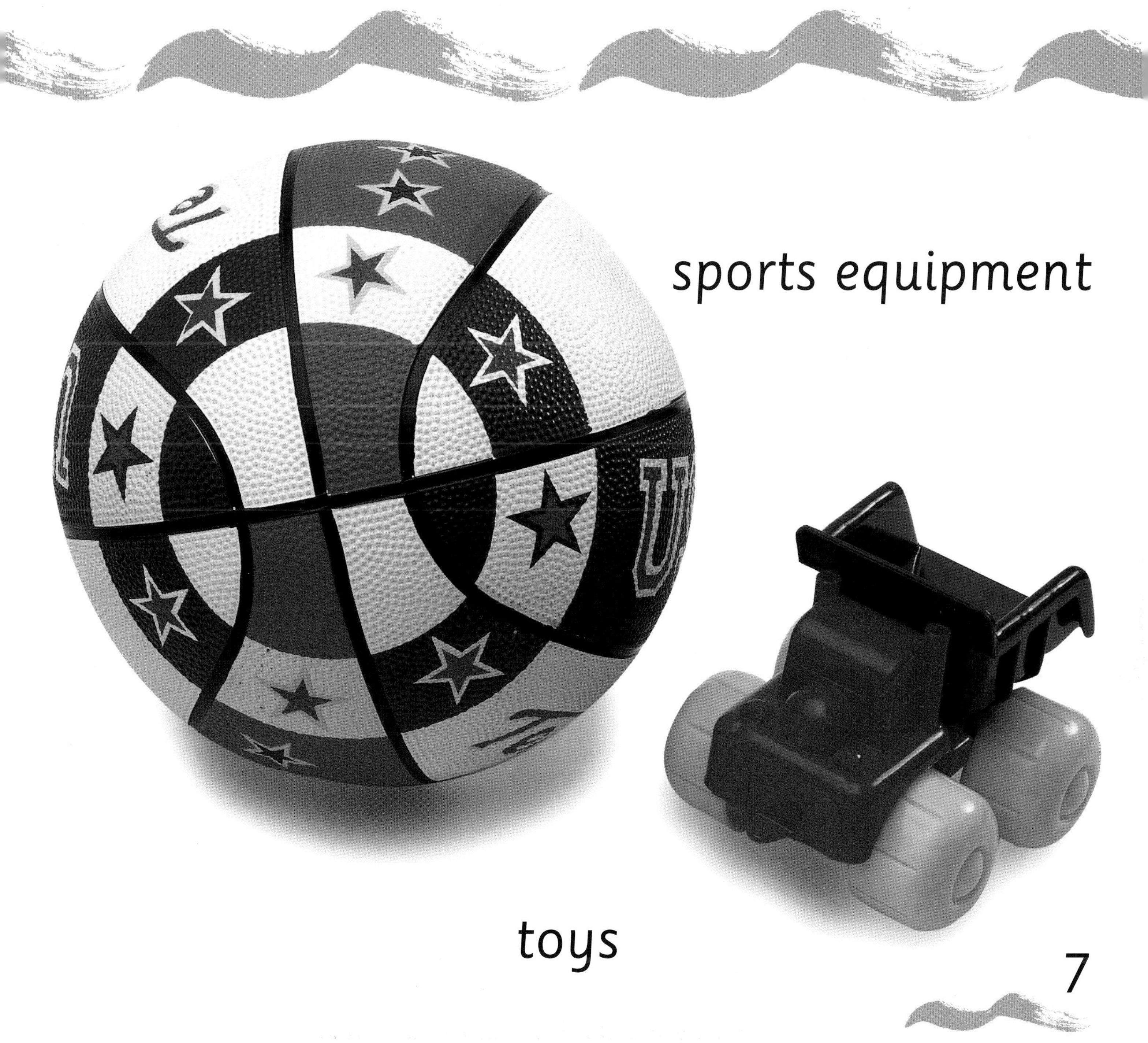

sports equipment

toys

# Food shops

These foods come from special types of food shops.

bread from the baker's

sausages from the butcher's

fish from the
fishmonger's

carrots from the
greengrocer's

# Transport

There are many different ways to travel around towns.

You need to
buy a ticket to
travel on a bus
or a tram.

# Wildlife

Wild animals can be found even in the centre of towns.

mouse

fox

beetle

pigeon

# Fast food

People in towns often want their food in a hurry.

Food that is served quickly is sometimes called fast food.

What type of fast food do you like?

# Play park

Many towns have play parks.

monkey bars

slide

swing

climbing frame

# Signs

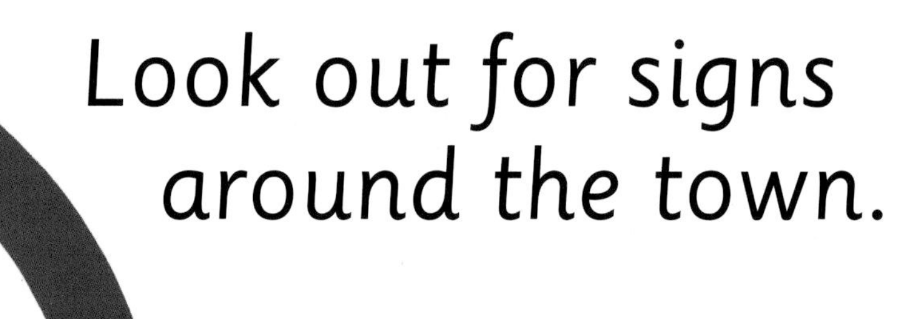

Look out for signs
around the town.

What do you think
these signs mean?

# Places to visit

There are many places to visit in a town.

theatre

library

market

museum

# Collections

*If you visit a town,*
*you could collect some*
*things to remind you*
*of your visit.*

FRESHLY BAKED
PIZZAS
TAKE AWAY MENU
ENGLAND'S
BEST
HISTORIC
PARKING VOUCH
No. C 725863
MINUTES
HOUR
YOUR TURN IS
E 44
To Lewes and Newhaven
DITCHLING ROAD
LEWES ROAD
LONDON ROAD
OPEN MARKET
LONDON RD
BRIGHTON RAIL STATION
TRAFALGAR STREET
NORTH LAINE
GLOUCESTER ROAD
NORTH ROAD
PRINCE REGENT SWIMMING COMPLEX
POLICE STATION
EDWARD STREET
QUEENS ROAD
CHURCH STREET
BRIGHTON MUSEUM
DOME
ROYAL PAVILION
ST JAMES' STREET
NORTH STREET
OLD STEINE
MARINE PARADE
MADEIRA DRIVE
To Brighton Marina
NORTH ST
DUKE ST
CHURCHILL SQ.
TOWN HALL
THE LANES
COACH STATION
GRAND JUNCTION ROAD
Palace Pier
WEST STREET
BRIGHTON CENTRE

# Glossary

**baker** someone who bakes bread and cakes
**butcher** someone who sells meat
**fishmonger** someone who sells fish
**greengrocer** someone who sells fruit and vegetables
**monkey bars** a set of bars that you swing along
**tram** rather like a bus that runs on rails
**transport** ways of travelling

# Index


animals 14–15
buildings 4–5
buses 11
collections 22–23
fast food 12–13
food 8–9, 12–13
libraries 20
markets 21
museums 21
play parks 16–17
shops 5, 6–7, 8–9
signs 18–19
theatres 20
trams 11
transport 10–11
wildlife 14–15